The Actin Cytoskeleton in Cell Motility, Cancer, and Infection

Colloquium Series on the Cell Biology of Medicine

Editor
Joel D. Pardee, Weill Cornell Medical College

Published

The Actin Cytoskeleton in Cell Motility, Cancer, and Infection
Joel D. Pardee
2010

Skeletal Muscle & Muscular Dystrophy: A Visual Approach
Donald A. Fischman
2009

How the Heart Develops: A Visual Approach
Donald A. Fischman
2009

Forthcoming Titles:

The Body Plan: How Structure Creates Function
Joel D. Pardee
2009

Bones: Growth, Strength, and Osteoporosis
Michelle Fuortes
2009

The Actin Cytoskeleton in Cell Motility, Cancer, and Infection
Joel D. Pardee
www.morganclaypool.com

ISBN: 9781615040063   paperback

ISBN: 9781615040070   ebook

DOI: 10.4199/C00003ED1V01Y200907CBM003

A Publication in the Morgan & Claypool Life Sciences series

COLLOQUIUM SERIES ON THE CELL BIOLOGY OF MEDICINE

Book # 3

Series Editor: Joel D. Pardee, Weill Cornell Medical College

Series ISSN Pending

The Actin Cytoskeleton in Cell Motility, Cancer, and Infection

Joel D. Pardee
Weill Cornell Medical College

COLLOQUIUM SERIES ON THE CELL BIOLOGY OF MEDICINE # 3

MORGAN & CLAYPOOL LIFE SCIENCES

ABSTRACT

By now it is abundantly clear that the cells of our bodies have the ability to move because of their actin cytoskeleton. From the earliest cell migrations in the embryo that serve to form the primordial tissue layers to the outgrowth of neurons, the contractions of our heart and skeletal muscles, and the metastasis of cancer cells, it is the regulated action of the actin cytoskeleton that propels our various motions. In this chapter we will examine the biochemical and cell biological mechanisms that control actin-based cell motility. We will see that cell migration along a substratum occurs in three fundamental phases that are exquisitely orchestrated in sequence. Cell migration occurs by an initial extension of the front edge of the cell, followed closely by contraction of the cell's rear end to push cytoplasm into the newly formed extension. As the cell's new leading edge extends, attachment sites are assembled on the cell floor, and extend through the plasma membrane to couple the migrating cell to the underlying surface. How the cell knows where to go, when to detach and re-attach during the migration, and how to keep each process under strict control is a story well worth hearing, because it is motility and adhesion that make a multicellular life such as ours possible.

ATTENTION READERS OF THE PAPERBACK VERSION: To view the video files associated with the digital version of *The Actin Cytoskeleton in Cell Motility, Cancer, and Infection*, please use the following URL: http://www.morganclaypool.com/r/actin

KEYWORDS

actin, cytoskeleton, motility, cancer, infection, protein polymers, enzyme complexes, organelles, vesicles, migration, microtubules, intermediate filaments, cilia, flagella, mitotic spindle, chromosomes, epithelial cells, muscle contraction, cytoplasm, cytokinesis

Contents

The Cytoskeleton

The cell is no longer considered to be a bag full of enzymes dissolved in a liquid cytoplasm. It is now known that the cytoplasm is an exquisitely ordered structure of properly placed organelles and enzyme complexes that are suspended from an intricate network of structural protein polymers termed the *cytoskeleton* (Figure 1). All movement of organelles and vesicles within the cell is regulated by this cytoskeleton, and it is clear that the cytoskeleton is responsible for all of the cell's external movement as well. In this lecture, we will consider how the cytoskeleton elicits cell migration.

The three main elements of the cytoskeleton are microtubules, intermediate filaments, and actin filaments (Figure 2). Microtubules are essential for (a) intracellular transport within the cytoplasm and transport between the nucleus and cytoplasm, (b) the structure and movement of all cilia and flagella, and (c) the structure of the mitotic spindle and movement of chromosomes on the spindle during cell division. Intermediate filaments give structural integrity to virtually all cells and tissues by providing an intracellular network of flexible cables that strengthen internal cell structure and stabilize cell-to-cell adhesion. It is this intercellular binding property that stably joins epithelial cells together to provide the protective functions of skin and the integrity of the intestinal mucosa. Actin is a highly conserved protein ubiquitous to all eukaryotic cells. Actin is absolutely required for (a) cell migration (Figure 3), (b) the contraction of muscle (both striated and smooth), (c) the structure and function of many cell protrusions (e.g., microvilli, filopodia, lamellopodia, blood platelet projections) (Figure 4), (d) division of the cytoplasm (cytokinesis) during telophase of cell mitosis (Figure 5), and (e) movement and placement of organelles within the cell. Actin filaments are also called *thin filaments* because of their very slender (70 Å) diameter.

WHY STUDY CELL MOTILITY?

The migration of cells is absolutely essential for many life processes. Neutrophils, macrophages, fibroblasts, nerve cell axons, amoebae, and embryonic cells all function through their ability to move. Embryonic cells must migrate within the embryo for development to proceed. Neuronal extensions (axons and dendrites) must migrate to and contact target cells in order to establish functional nerve conduction pathways. In fully differentiated animals, a number of migratory cell types sustain the

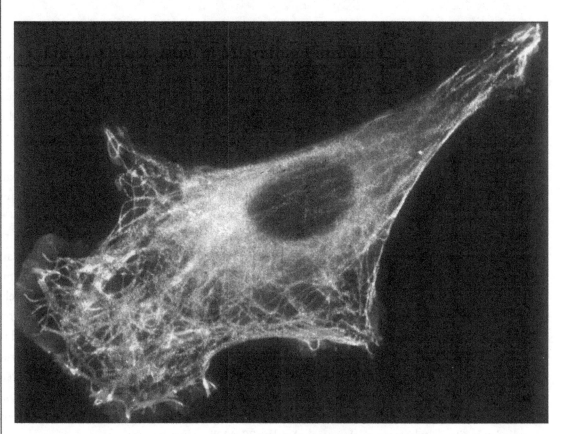

FIGURE 1: Cytoskeleton of migrating fibroblast. (Reproduced from Darnell, et al. (1986), Cover, Mol. Cell Biol.: Feeling/Organism: 1st edition with permission of W.H. Freeman). Courtesy of J. Victor Small.

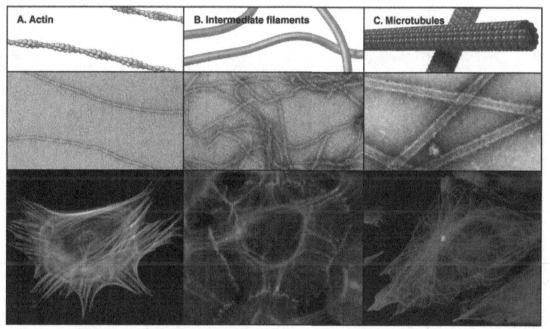

FIGURE 2: The cytoskeletal polymers: (A) Actin filaments, (B) Intermediate filaments, (C) Micro-tubules. The top panels illustrate scale drawings. The middle panels show electron micrographs of negatively stained specimens of each polymer. The bottom panels show fluorescence light micrographs of cultured cells stained for each type of cytoskeletal polymer. (Reproduced from Pollard and Earnshaw: Cell Biology, www.studentconsult.com with permission of Elsevier Ltd.)

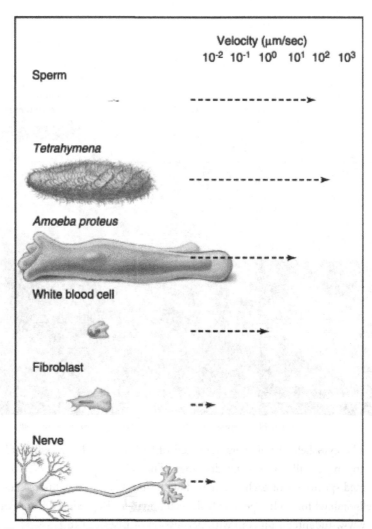

FIGURE 3: Rates of cell migration. A nerve or a fibroblast moves ~0.1 mm/day and a leukocyte ~100X faster. (Reproduced from Pollard and Earnshaw: Cell Biology, www.studentconsult.com with permission of Elsevier Ltd.)

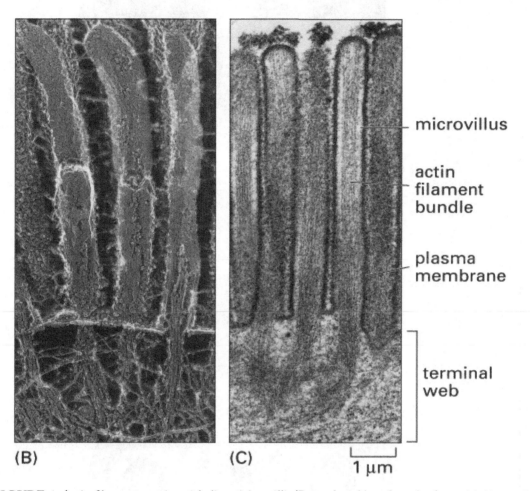

(B) (C)

microvillus

actin
filament
bundle

plasma
membrane

terminal
web

1 µm

FIGURE 4: Actin filament core in epithelium microvilli. (Reproduced based on the figure 16-41 part 2 of 2 from Molecular Biology of the Cell, 4th ed. with permission of Garland)

(a)

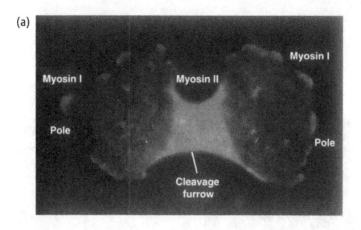

(b)

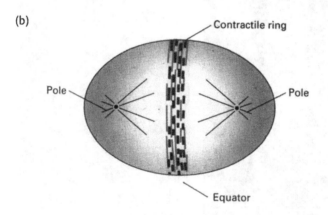

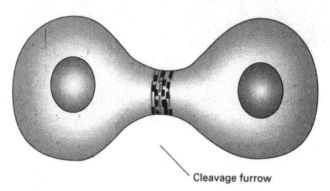

FIGURE 5: Cytokinesis: actin and myosin-II mediated contraction. Micrograph courtesy of Yoshio Fukui.

adult organism. Fibroblasts continuously move through the connective tissues to lay down the extracellular matrix, leukocytes migrate out of the bone marrow into the bloodstream, and thence to surrounding connective tissue to combat infection, and blood cell precursors in the bone marrow migrate to enter the circulation. Consequently, regulating cell motility has become of interest in to modern biomedical researchers and medical practitioners.

CELL MOTILITY IN CANCER

Most cancer patients do not die of their primary tumor. In fact, when the initially transformed and proliferating cells remain in their tissue of origin, the cancer is not considered a malignancy, but rather a benign tumor. The cancer killers are metastatic cells that escape the primary tumor, enter the blood vessels and lymphatic ducts, disseminate throughout the body, and seed distant tissues with micrometastases (Figure 6). Subsequent growth of these single cell seeds into multicellular tumors disrupts tissue and organ function, drains large amounts of energy to feed their growth, and finally metabolically starve the body to death (termed *cachexia*). As a result of transformation, normal sessile cells gain the ability to proliferate and migrate. Carcinomas, which account for about 80% of all cancers, are malignancies that derive from epithelial cells that can proliferate and migrate through the basement membrane to gain access to the vasculature (Figure 7). It is important to realize that metastatic cells are not simply shed into the surrounding connective tissue due to the pressure of proliferation, but instead actively create a hole in the basement membrane and proceed to crawl into and through the underlying connective tissue. The process is repeated when the cell reaches the vasculature, which is itself coated with a basement membrane, and the metastatic cell squeezes through the tight junction that binds vascular endothelial cells together. Cell motility, resulting from dynamic rearrangement of the cancer cell actin cytoskeleton, is therefore the driving force of metastasis.

CELL MOTILITY IN INFECTION

Another critical role for actin-driven cell motility occurs in response to infection. It is a dirty world out there, so whenever pathogens breach an epithelium, often through a puncture or laceration, the body fights the ensuing infection by mounting an acute inflammatory response (Figure 8). Redness, swelling, pain, heat, and immobilization of the affected area follow. These cardinal signs of infection have been recognized for 2,000 years, since the time of the Greek physician, Galen. To effectively neutralize the invading pathogen, resident and plasma leukocytes must be attracted to the infected connective tissue. When injury occurs, the bacteria shed endotoxins (a substance from the cell coat recognized as foreign), which diffuses through the local connective tissue. As endotoxin reaches the endothelial cells of postcapillary venules of capillary beds, it initially triggers a contraction of

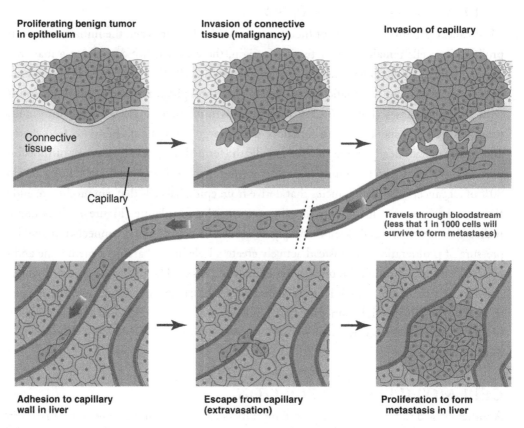

FIGURE 6: Tumor metastasis. Migration of cancer cells from epithelium. (Based on the figure 23-15 from Molecular Biology of the Cell, 4th ed./Garland).

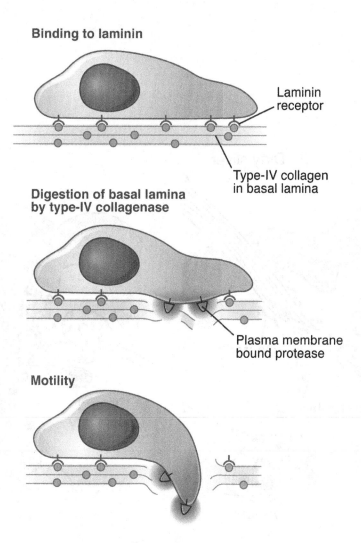

FIGURE 7: Invasion across the basement membrane. (Based on the figure 24-18 from Molecular Biology of the Cell, 3rd ed./Garland)

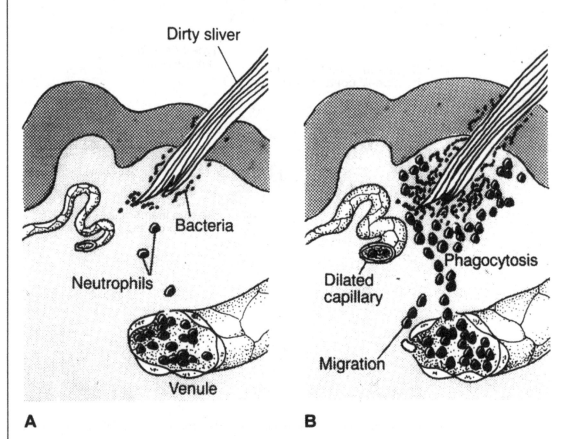

FIGURE 8: Puncture of skin initiating inflammation.

smooth muscle around the venule. This functions to stop serious bleeding and allows a blood clot to form. The endotoxin secondarily stimulates mast cells in the connective tissue to explosively degranulate, releasing histamine and heparin. Histamine causes the endothelium of vessels to become more permeable and heparin retards blood clotting. This leads to a localized influx of plasma into the infection site (edema), causing swelling, redness (due to an increased local concentration of red blood cells) and pain (from increased pressure on sensory nerves). The result of edema is a slowing of circulation in the area, facilitating migration of white blood cells from blood into the connective tissue.

The migration of leukocytes from blood vessels is called *diapedesis* (Figure 9). Endotoxin diffusing from the infectious agent contacts endothelial cells lining blood vessels causing insertion of a cell surface receptor called *LECAM* (leukocyte endothelial cell adhesion molecule) on the apical plasma membrane facing the blood vessel lumen. Passing leukocytes attach weakly to LECAM and begin to roll along the endothelial surface giving rise to margination of the white blood cell. Upon margination, leucocytes attach more tightly to the vessel endothelium exactly at the site of the infection. The leukocytes then break tight junctions between endothelial cells, insert a pseudopod into the gap, and migrate into the connective tissue containing the proliferating pathogen.

ADHESION AND TRANSMIGRATION SIGNALING

It is now clear that leukocyte adhesion and transmigration are determined largely by the binding of complementary adhesion receptor molecules on the leukocyte and endothelial cell surfaces and their chemical mediators (Figure 10). Chemoattractants and cytokines control these processes by modulating the surface expression and/or avidity of cell adhesion molecules. The principle adhesion receptors used, in order of their appearance in the process, are:

1. Selectins that bind through their lectin domain to sialylated oligosaccharides of glycoproteins and glycolipids of the glycocalyx of the endothelium. This gives rise to rolling adhesion and margination of the leucocyte.
2. Integrins are transmembrane glycoproteins, made up of α and β chains that bind to leukocyte function-associated antigen expressed on the endothelial cell surface. Binding of integrin results in transmembrane activation of the actin cytoskeleton of the leukocyte, resulting in (a) tight binding to endothelium and (b) expression of contractile proteins in the leukocytes generating the mechanical forces underlying cell motility.
3. PECAM is a high-affinity homodimer receptor expressed both at the endothelial cell junctional complex and on the leukocyte. Extremely tight homodimer interactions between these two cell types allow transmigration across the tight junctions of the endothelium.

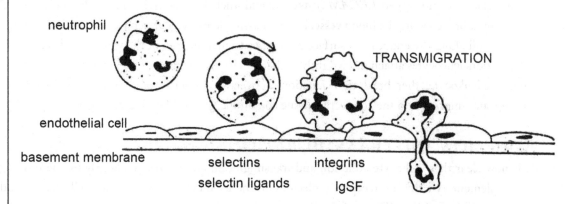

FIGURE 9: Neutrophil transmigration.

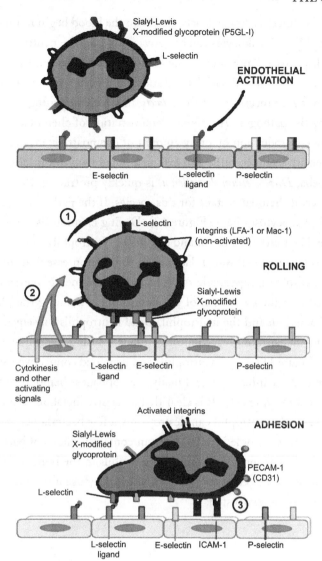

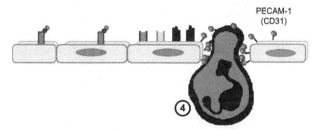

FIGURE 10: Adhesion receptors involved in leukocyte transmigration.

Once in the connective tissue, leucocytes from the blood begin to migrate toward the pathogen target (Figure 11). All leukocytes can be activated to migrate: neutrophils phagocytose bacteria, eosinophils attack parasites, monocytes convert to macrophages once in the connective tissue to phagocytose bacterial cell and tissue debris, and lymphocytes mount an immune response. The leucocytes migrate by a process called *chemotaxis*, in which migrating cells follow a trail of small molecules shed by the pathogen. A striking demonstration of chemotactic stimulation is depicted in Figure 12, where introduction of a tiny amount of chemoattractant quickly activates the side of the cell nearest the attractant to ruffle and extend a pseudopod toward the target. In this particular example, an amoeba, *Dictyostelium discoideum*, is quickly protruding toward a micropipette exuding cAMP, a powerful chemoattractant for this species. If the pathogen is a bacterium, neutrophils chemotax to and phagocytose them (Figure 13). During chemotaxis, the migrating cell extends a pseudopod in the direction of the target and engulfs it. Neutrophils contain digestive enzymes such as lysozyme (lyses bacterial cell walls), lactoferrin (binds iron essential for bacterial metabolism), collagenase, acid hydrolases, superoxide radical (lyse plasma membrane), and peroxides. These enzymes are released into the cytoplasm of the neutrophil containing the phagocytosed bacterium, killing both the bacterium and the neutrophil. Dead neutrophils subsequently release their hydrolytic enzymes into the surrounding connective tissue, causing hydrolysis and necrosis of healthy tissue in the area (Figure 11). Dead neutrophils and degraded connective tissue constitute the commonly observed pus at the infection site. Finally, macrophages chemotax to the area to phagocytose the detritus of dead and dying cells. It is clear that a massive metabolic effort is required to neutralize an infection. A single neutrophil can destroy only a few bacteria before it kills itself by its own digestive enzymes. Since a single infected site can contain millions of bacteria, the expenditure of ATP energy required to generate an army of neutrophils in the bone marrow, diapedese from the vasculature, chemotax to the target bacterium, and phagocytose it is truly impressive. It is no wonder that infections leave us exhausted.

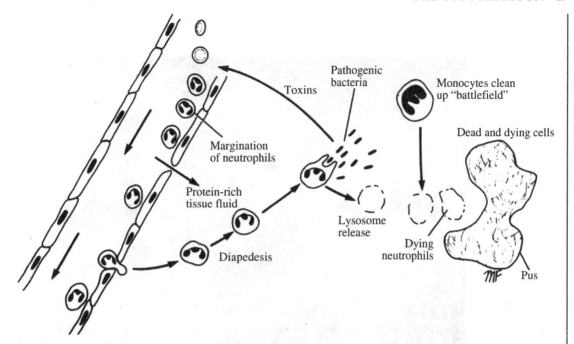

FIGURE 11: Diapedesis.

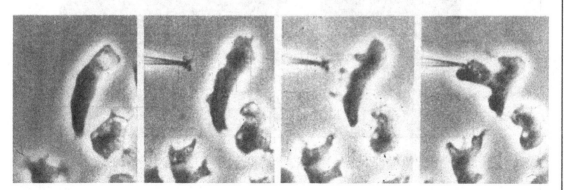

FIGURE 12: Amoeba chemotaxis. Courtesy of Gunter Gerish.

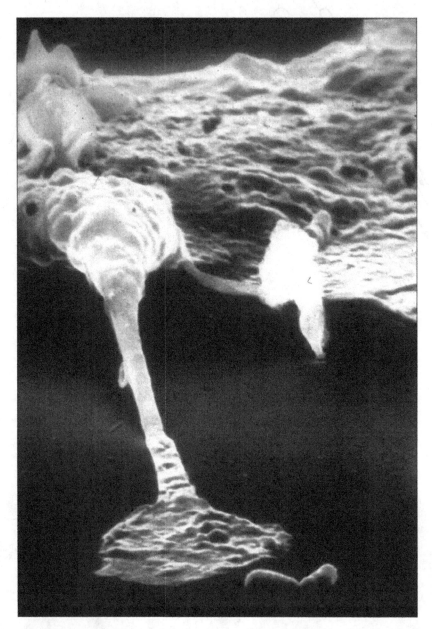

FIGURE 13: Neutrophil migration toward a bacterium. Courtesy of Thomas Stossel.

How Cells Move

A series of cell biological experiments performed by Malawista and his colleagues approximately 20 years ago focused attention on the essential role of the actin cytoskeleton in cell migration (Figure 14).

1. Neutrophils were allowed to adhere to a substratum containing collagen, and the temperature was then raised to 40°C. This caused the plasma membrane beneath the nucleus to adhere very tightly to the substratum (Figure 14a,b) but allowed the rest of the cell free to move (Figure 14d–f).
2. Opsonized red blood cells (red blood cells coated to facilitate their phagocytosis by neutrophils) were then added to the neutrophil culture. The neutrophils extended a cytoplasmic process (pseudopod) toward the red blood cells in a process termed *chemotaxis*, leaving the adherent nucleus behind (Figure 14g–i). This migration continued to the point at which the nucleus remained attached only by a long tether to the cytoplasmic pseudopod. Finally, migration proceeded so far from the attached nucleus that the tether snapped, leaving the cytokinoplast (literally a piece of moving cell) (Figure 14j).
3. Amazingly, the cytokinoplast retains the ability to migrate and search out and phagocytose the target cell in an attempt to destroy it (Figure 14k,i). If another cell or cytokinoplast gets to the target first, the cytokinoplast can even divert its path to a new target.

When the contents of the cytokinoplast were examined by scanning electron microscopy (Figure 15), the fascinating observation was that it consisted merely of a plasma membrane and its associated cortical actin filament network. It contained no organelles, microtubules, intermediate filaments, or nucleus. However, it did contain the target cell, proof that the cytokinoplast could phagocytose. Because the cytokinoplast lacked lysosomes, it could not digest the phagocytosed target cell, and because it lacked mitochondria (and therefore a renewable source of ATP), the cytokinoplast was short lived.

The conclusion of these experiments is that all of the machinery for cell motility and chemotaxis is confined to the region of the plasma membrane and its associated actin filament network. Thus, motility processes are not directly regulated by changes in nuclear gene expression, since the cytokinoplast lacks a nucleus.

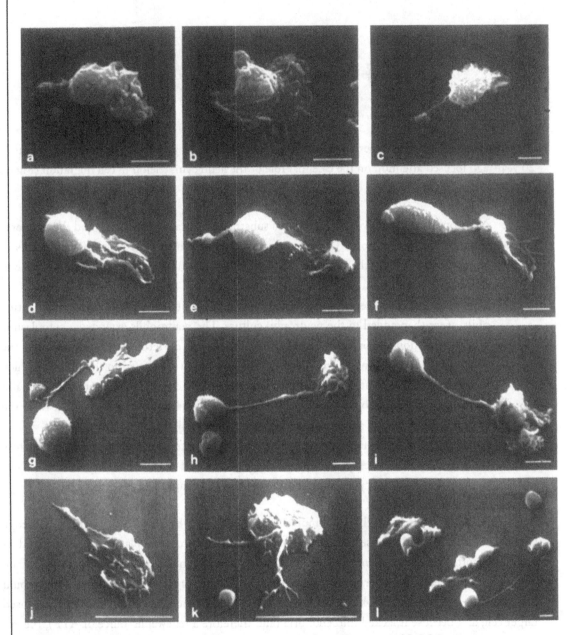

FIGURE 14: Malawista experiment creating cytokinoplasts. Courtesy of S. Malawista.

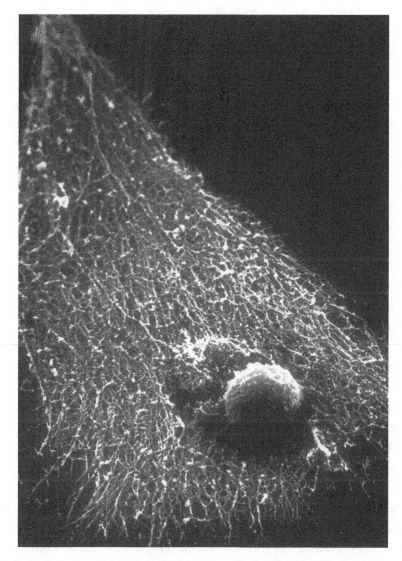

FIGURE 15: Actin filament network in a cytokinoplast. Courtesy of S. Malawista.

Regulation of the Actin Cytoskeleton

Detailed microscopic examination of migrating cells reveals that cells move forward in three distinct steps: (1) extension of the leading edge, (2) adhesion of the new extension to the substratum, and (3) retraction of the lagging edge of the cell (Figure 16). Extension requires rearrangements of the actin filaments beneath the plasma membrane. Adhesion occurs by the formation of attachment plaques on the cell floor. Retraction is a pinching off of the tail of the cell by actomyosin contraction. The orchestration of these steps into concerted cell migration is beautifully visualized in the movie made by Dr. Vic Small presented in Video 1.

EXTENSION OF THE LEADING CELL EDGE

There are three basic types of extensions that can occur at the leading edge of a migrating cell: pseudopodia, lamellipodia, and filipodia. Pseudopodia are bulbous extensions that appear to bubble out from the leading edge of the cell during chemotaxis (Figures 12 and 13). Lamellipodia are also called *ruffling edges* because the leading edges of the cells are constantly forming and ruffling backwards over the top cell surface (Figure 17). Filipodia are needlelike projections spiking out from the dorsal surface of the cell. Strong evidence indicates that the major force underlying all of these extensions is the forward polymerization of actin filaments. To understand the mechanisms underlying directed polymerization, we must examine some of the biochemical and cell biological properties of actin.

ACTIN FILAMENT STRUCTURE AND FUNCTION

Humans have six genes encoding different isoforms of actin. The proteins are remarkably conserved (~90% conservation) across the eukaryotic kingdom. In our bodies, there are three forms of actin: termed α, β, and γ. α-Actin is restricted to muscle cells, while β- and γ-actins are found in virtually all nonmuscle cells. It is currently believed that virtually the whole surface of the actin molecule is involved in essential functions—binding to other actin molecules, to a large family of actin associated proteins, to myosin family members, or to other cytoskeletal proteins involved in a wide variety of intracellular functions.

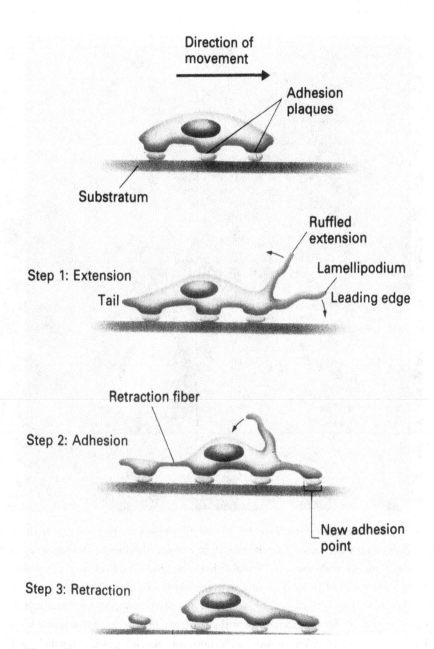

FIGURE 16: How cells move.

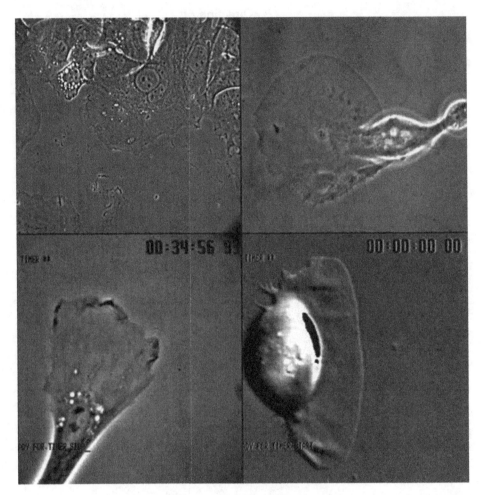

VIDEO 1: Cells moving in vitro. Top left: Mouse fibroblasts moving into an artificial wound created in a Petri dish (total video time, 3 h). Bottom left: a chick fibroblasts, moving alone (total video time, 2 h). Top right: mouse melanoma cell (total video time, 20 min). Bottom right: trout epidermal keratocyte (total video time, 4 min). The images were recorded using either phase contrast optics (chick fibroblasts and mouse melanoma cell) or Nomarski interference optics (fish keratocyte). The differences in migration speed can be appreciated from the different durations of the movie sequences. Courtesy of J. Victor Small. (See http://www.morganclaypool.com/userimages/ContentEditor/1258743611322/Video1-fig17.mov).

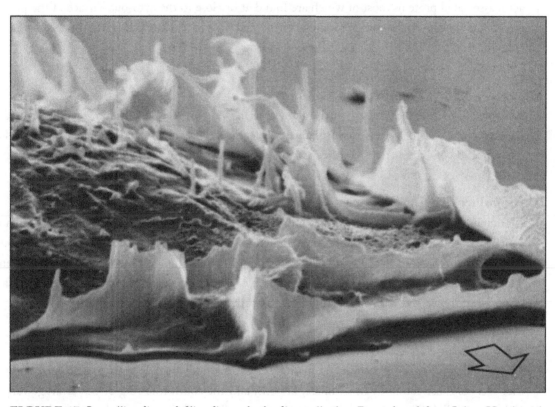

FIGURE 17: Lamellipodia and filipodia at the leading cell edge. Reproduced from Julian Heath with permission from Garland.

Actin filaments are helical structures approximately 5 nm in diameter and are composed of 42-kDa globular actin (G-actin) subunits (Figure 18). Each filament has a fast growing end (the plus end) and a slow growing end (the minus end). Many years of research have shown that actin itself is not very important in regulating its assembly into filaments. Under physiological conditions of salt, Mg^{2+}, and ATP, >98% of G-actin will spontaneously assemble into stable filaments (Figure 19) that cannot be disassembled or rearranged into networks, bundles, or fragments. Instead, it appears that changes in the architecture of the filament network are manipulated by a large number of actin associated proteins most of which are found at or close to the cytosolic surface of the plasma membrane. These actin binding proteins act in several ways to regulate the aggregation state of actin (Figure 20).

Actin Severing Proteins

At high concentrations of Ca^{2+} (μM), actin severing proteins (gelsolin, villin, severin, and actin depolymerizing factor) cleave actin filaments into shorter fragments and cap their ends. At low concentration of Ca^{2+} ($<10^{-7}$ M), they act to nucleate new sites of filament formation.

Monomer Binding Proteins

Monomer binding proteins (thymosin and profilin) bind G-actin and form an actin–profilin or actin–thymosin complex. The thymosin–actin complex inhibits plus end growth. Profilin competes with thymosin for free actin subunits, and when profilin binds, monomer actins add to the plus end. The reverse is true for thymosin. Phosphorylation of profilin regulates its activity and is especially important for plus end growth of actin filaments at the plasma membrane.

Cross-Linking Proteins

Cross-linking proteins (α-actinin and filamin) cross-link individual fibers in parallel to form bundles such as those found in stress fibers, filipods, and the contractile ring of the cleavage furrow. Others cross-link filaments at right angles (orthogonally) to form the complex networks found in the cell cortex beneath the plasma membrane.

Capping Proteins

Capping proteins (capZ and tropomodulin) stabilize filaments by binding to free ends. They are often found in muscle, where they attach actin filaments to the Z-line (capZ) and to other structural proteins (tropomodulin).

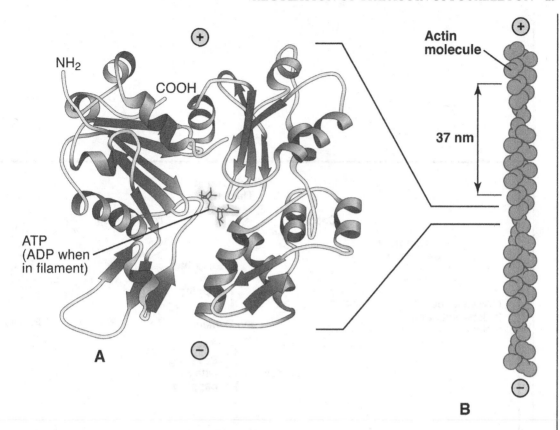

FIGURE 18: Structure of actin. (Based on the figure 16-7 from Molecular Biology of the Cell, 4th ed./Garland)

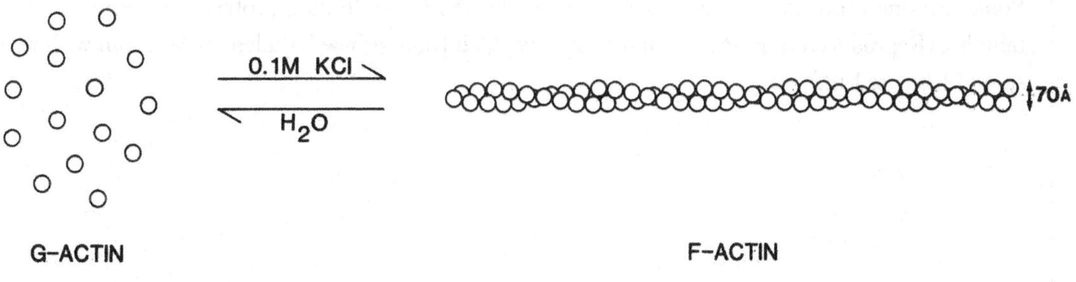

FIGURE 19: Polymerization of actin.

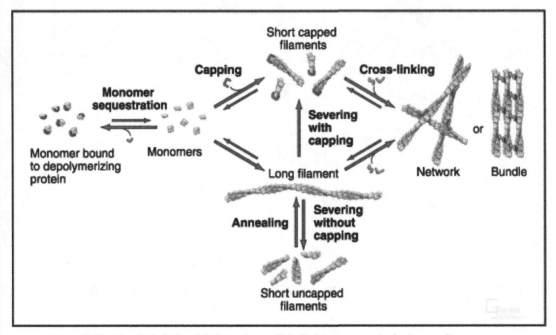

FIGURE 20: Families of actin-binding proteins. Monomer-binding proteins favor either ATP-actin or ADP-actin. Capping proteins bind specifically to either the barbed end or pointed end of filaments. Some severing proteins also cap, whereas others do not. Cross-linking proteins can form networks or bundles. (Reproduced from Pollard and Earnshaw: Cell Biology, www.studentconsult.com with permission of Elsevier Ltd.)

Nucleating Proteins

Nucleating proteins (actin-related protein [ARP]) are an important class of proteins that nucleate actin filaments near the plasma membrane (Figure 21). The main player is the ARP2/3 complex. It can bind at the plus end to initiate addition of monomers to the end of the filament. ARP2/3 complex can also bind along the shaft of an actin filament to initiate branching of actin filaments. This leads to web formation in the extending edge of the cell.

THE ACTIN ASSEMBLY MODEL FOR PSEUDOPOD EXTENSION

In this model, protrusive force is generated by the assembly of actin filaments against the plasma membrane, forcing extension (Figure 22). Binding of chemoattractants to a receptor at the leading cell edge triggers a series of actin disassembly, uncapping, and reassembly events that cause plasma membrane extension. Steps in the postulated process for which there is experimental evidence include:

1. Binding of extracellular chemoattractant to a G-protein receptor on the plasma membrane releases intracellular stores of Ca^{2+}, triggering momentary fragmentation of long actin filaments present in the cell cortex. The resulting fragments are capped to prevent reassembly. In general, elevated $[Ca^{2+}]$ in the cytosol activates actin fragmentation and disassembly, while removal of Ca^{2+} causes actin reassembly, cross-linking, and bundle formation.

2. Actin monomers bound by escort proteins such as profilin are transported from the center of the cell to the cortex, where actin is released for assembly into filaments.

3. As Ca^{2+} levels decrease to prestimulation levels, actin assembly is initiated on capped fragments (Figure 21). Caps are dissociated from ends of actin fragments, escort proteins are dissociated from monomeric actin subunits, and net assembly off the fragment ends is initiated.

4. Growth of filaments occurs synchronously with cross-linking of filaments into rigid networks. Actin assembly against the plasma membrane distends the cell in the direction of the chemotactic gradient.

5. Activation of myosin I (tail-less myosin) is thought to slide newly formed actin filaments past one another and/or along the plasma membrane, further enhancing pseudopod extension.

6. Contraction of actin cables at the more posterior floor of the cell occurs by a myosin II mediated process that brings the rear of the cell forward (Figure 23).

As mentioned earlier, cell migration facilitates metastasis of cancer cells. The actual movement of a melanoma cancer cell is shown in Video 2, where extension of the leading edge of the

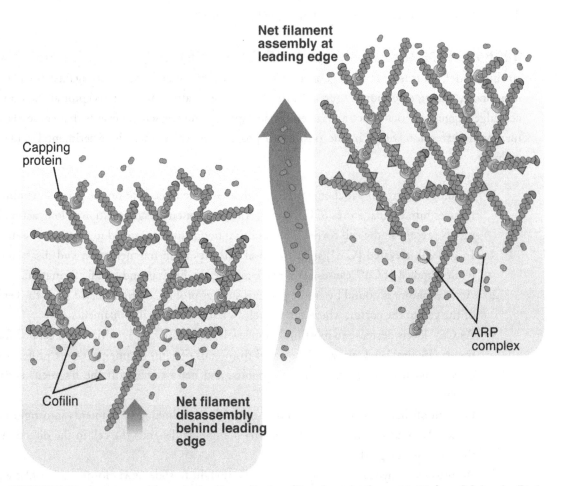

FIGURE 21: Actin assembly at the leading cell edge. (Based on the figure 16-90 from Molecular Biology of the Cell, 4th ed./Garland)

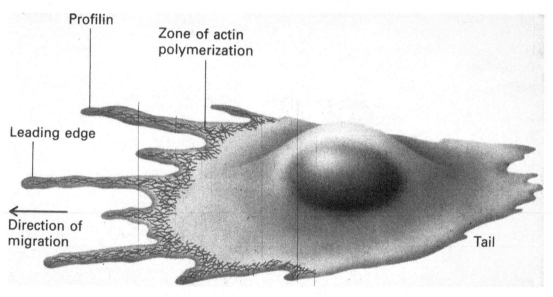

FIGURE 22: Extension by assembly of actin filaments.

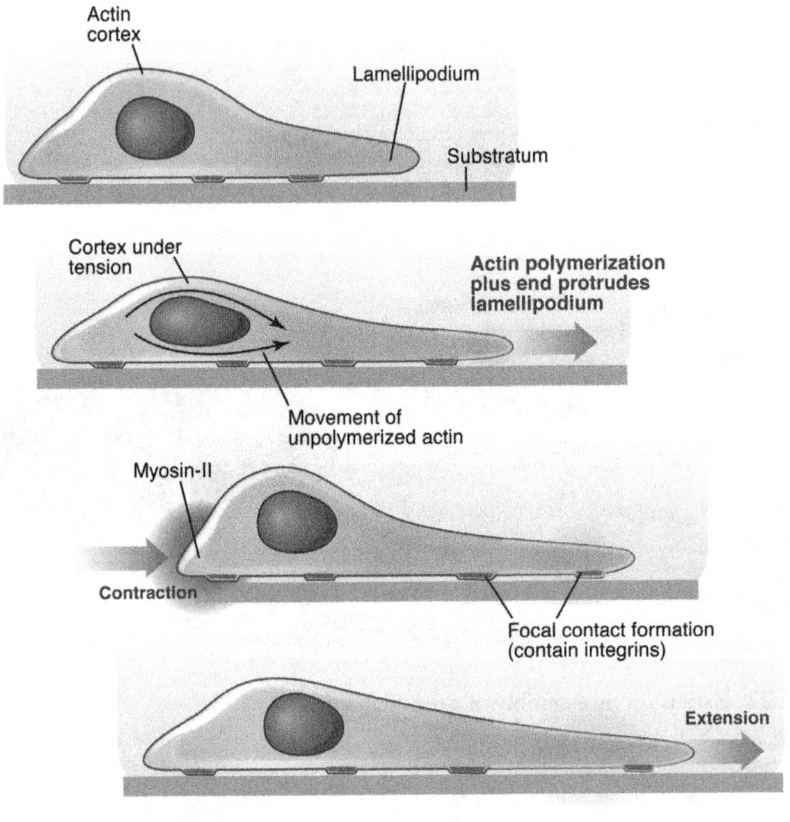

FIGURE 23: Stages of cell migration. (Based on the figure 16-85 from Molecular Biology of the Cell, 4th ed./Garland)

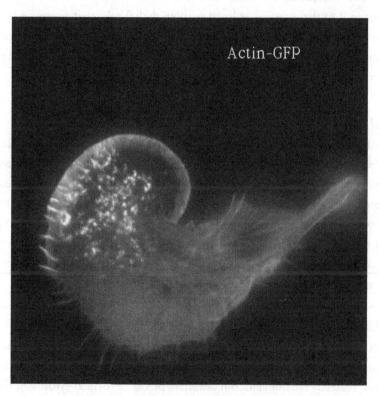

VIDEO 2: The actin cytoskeleton in a migrating melanoma cancer cell. Actin was tagged with green fluorescent protein (actin-GFP) to visualize the rearranging actin. Courtesy of J. Victor Small. (See http://www.morganclaypool.com/userimages/ContentEditor/1258743817358/Video2-fig23.mov).

cell by actin polymerization is followed by retraction at the rear of the cell occasioned by actomyosin contraction. Extension distorts the plasma membrane forward and retraction pushes the cell contents into the newly created forward space.

THE ROLE OF ACTIN IN INTESTINAL ABSORPTION

Absorption of nutrients in the small intestine and resorption of water by the large intestine are accomplished by presenting an enormous surface area to the gut contents. To achieve this critical property, the wall of the intestine is lined with epithelial cells that possess a brush border on their surface that protrudes into the lumen of the intestine (Figure 24). This brush border of innumerable microvilli projecting from the surface of the absorptive epithelial cells lining the gut creates a surface area the size of a tennis court to quickly absorb nutrients and water.

Intestinal Microvilli

The core of each microvillus is composed of a very tightly organized bundle of actin filaments linked to the lateral plasma membrane by side arms of myosin I and calmodulin (Figure 25). At the tip, there is a dense staining plaque reminiscent of the Z-band of striated muscle. The plus ends of the filaments anchor into the distal tip. The actin filaments in the microvillus extend down into a terminal weblike structure that extends horizontally across the width of the epithelial cell just at the base of the microvilli. This web of actin bundles is contractile and contains myosin II filaments intermingled with the actin web. Contraction of the terminal web keeps the microvilli erect, thus presenting maximal surface area to the contents of the intestine.

THE ROLE OF ACTIN IN BLOOD CLOTTING

Clot formation in the blood requires a cascade of metabolic pathways, the end result of which is the activation of blood platelets to plug a wound or patch over a denuded section of the blood vessel endothelium. Platelets are not cells, but are small pieces of cortical cytoplasm that have pinched off a huge cell, the megakaryocyte. Consequently, like the cytokinoplast, they do not contain a nucleus, but they do possess a completely functional plasma membrane and cortical actin filament network. They also contain microtubules.

Platelet Activation

Upon activation, platelets quickly change their shape and adhesiveness (Figure 26). They convert from a perfectly round disk that does not adhere to the vessel endothelium to an irregular and spread in flat shape that tightly binds both the vessel endothelium and other activated platelets. The mor-

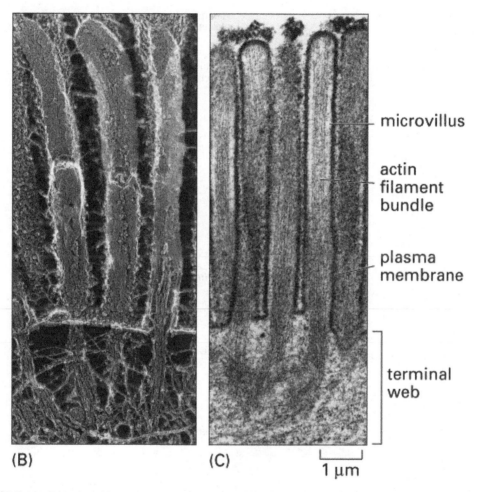

(B)   (C)   1 μm

FIGURE 24: Actin bundles in intestinal microvilli. (Reproduced based on the figure 16-41 part 2 of 2 from Molecular Biology of the Cell, 4th ed. with permission of Garland)

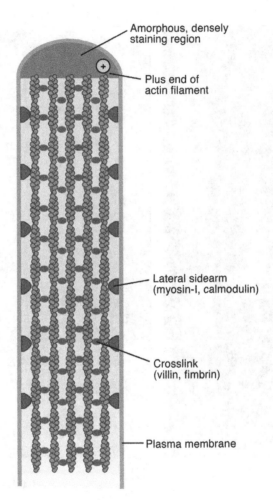

FIGURE 25: Cross-linked actin filaments in a microvillus. (Based on the figure 16-41 part 1 of 2 from Molecular Biology of the Cell, 4th ed./Garland)

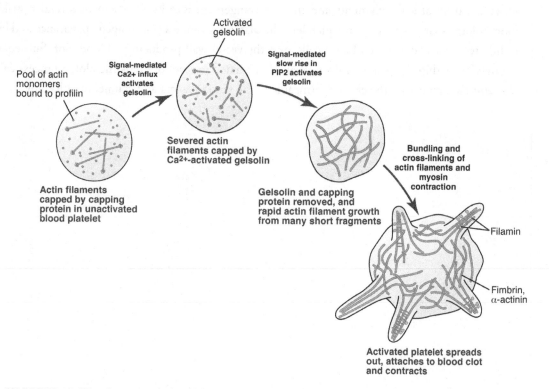

FIGURE 26: Platelet activation and contraction. (Based on the figure 16-47 part 1 of 2 from Molecular Biology of the Cell, 4th ed./Garland)

phologic and functional conversion of platelets during activation results from a dramatic rearrangement of its actin cytoskeleton. Circulating platelets contain large pools of actin monomers bound to profilin, together with actin filaments with capped ends. Upon activation of the platelets by a plasma membrane G-protein-mediated Ca^{2+} influx, there is an activation of gelsolin and a severing of the filaments. Secondarily, there is a slow rise of phosphatidyl inositol diphosphate, which initiates a release of gelsolin and actin capping protein from the end of the filament. Uncapped filaments undergo explosive actin polymerization. The actin filaments are then bundled and cross-linked. Polymerization of actin filaments and their rearrangement into bundles and nets render striking morphologic shape changes to the platelet. The platelets assume a star-shaped appearance and bind tightly to other platelets, red blood cells, and the vessel wall producing a blood clot. Subsequent contraction within the platelet by a myosin II-mediated process tightens the clot, a process called *clot retraction*, rendering the clot impermeable to the blood plasma and staunching bleeding.

The Role of Myosin in Cell Motility

The protein myosin is important in nonmuscle cell motility—predominantly, but not exclusively, in the contraction that occurs at the tail end of the migrating cell (seen in Figure 23). It is now known that are a large number of myosin isoforms encoded by a correspondingly large number of genes. Some are essential for muscle contraction (myosin II) and some are required nonmuscle cell motility (myosin I and II).

ACTOMYOSIN CONTRACTION BY MYOSIN II PUSHES THE CELL FORWARD

Following pseudopodial extension at the leading edge of the cell, a contractile event takes place at the rear of the cell that forces movement of the cell mass forward into the new pseudopodium. In a striking set of observations, Yoshio Fukui showed that myosin II is responsible for this rear end contraction and that myosin II is recruited to the rear of the cell when the cell is stimulated to move (Figure 27). Serial sections taken through a sessile (nonmotile) cell (Figure 27a–d) reveal that myosin II (white fluorescence) encircles the entire cell in a thin rim beneath the plasma membrane. However, when the cell is stimulated to move toward a chemotactic stimulus, virtually all of the myosin II moves to the back end of the cell (Figure 27i–k). Likewise, when the cell undergoes cell division, the cortically disposed myosin II all moves to the cleavage furrow dividing the two daughter cells (Figure 27e–h). This makes enormous sense, because it is precisely at these locations that the cell must undergo contraction to complete either a forward step during migration or cytokinesis during cell division. Remembering that it is myosin II that causes actin filaments to slide past the myosin thick filament in muscle to shorten the sarcomere during muscle contraction, the same mechanism with myosin II occurs to pinch the cell's rear, squeeze cytoplasm into the extending front, and force the cell's center of mass forward.

MYOSIN I HELPS EXTEND THE LEADING EDGE

However, another myosin isoform, myosin I, is at work to help extend the leading edge of the cell (Figure 28). Myosin I concentrates at the cell's leading edge at the same time that myosin II con-

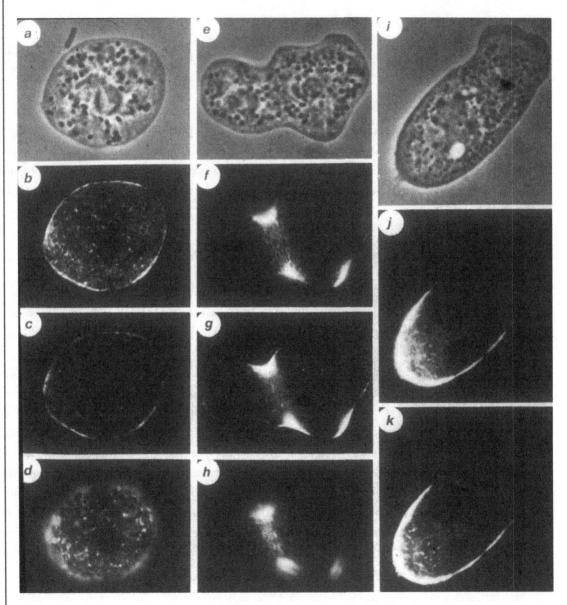

FIGURE 27: Myosin II in nonmotile (a), dividing (e), and motile (i) cells. Courtesy of Yoshio Fukui.

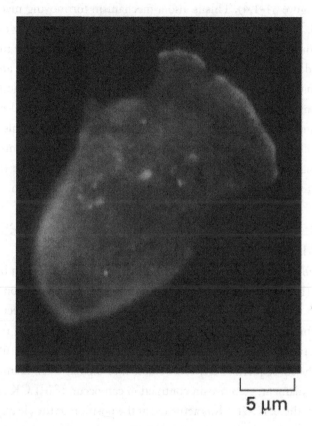

5 μm

FIGURE 28: Redistribution of myosin-I (green) to the leading edge and myosin-II (red) to the lagging edge of a migrating cell. Courtesy of Yoshio Fukui.

centrates at the lagging edge. This emphasizes the functional differences between myosin isoforms and can be explained by their strikingly different structures. Myosin II possesses a long tail (Figure 29), and myosin I is tail-less (Figure 30). Myosin II forms bipolar thick filaments by antiparallel bundling of it tails (Figure 31-3), and it is this bipolar arrangement that creates the sliding of actin filaments past each other to contract the cell region attached to it. Conversely, tail-less myosin I simply slides actin filaments along the plasma membrane (or against it) to cause expansion instead of contraction (Figure 31-1,4). This is also a mechanism for moving myosin molecules around in the cell (Figure 32), since tail-less myosins can slide along the relatively more fixed actin filaments in the cortical actin networks. However, the question remains as to how myosin II can move in the cortex. Its exposed tail should spontaneous assemble into thick filaments and result in contraction of the actin cortex instead of movement through it. Two unanswered questions arise: how does myosin II translocate from the cell cortex to the posterior edge of the cell, and how are actomyosin contractile fibers formed at the contraction site? Identical mechanisms are also presumably involved in formation of the cleavage furrow during mitosis. The precise mechanisms of myosin II movement and actomyosin fiber formation in nonmuscle cell are ongoing research topics, but both processes certainly involve the known highly regulated phosphorylation of myosin.

PHOSPHORYLATION OF MYOSIN II REGULATES TAIL ASSEMBLY

Depending on cell type, assembly of myosin II into thick filaments can be regulated by either tail phosphorylation (heavy-chain phosphorylation) or head phosphorylation (light-chain phosphorylation). For instance, during chemotaxis of neutrophils to the site of infection, a myosin light-chain kinase (MLCK) binds to the myosin light-chain subunit located in the myosin head and transfers the terminal phosphate from ATP to the light chain (Figure 33). This results in straightening of the myosin tail and the subsequent ability to assemble into bipolar thick filaments. With the formation of a bipolar thick filament, actomyosin contraction can occur. If MLCK acts at the rear of the cell, contraction occurs there. If MLCK is activated at the position of the cleavage furrow, cytokinesis via contraction occurs. It is presumed that it is the bent tail conformation of myosin that translocates about the cell cortex on actin filaments, since bent tail myosin can move motor along F-actin but cannot form a contractile complex. In vitro reconstitution experiments have shown that individual myosin monomers can rapidly move along actin filaments, that LC phosphorylation allows reassembly into bipolar thick filaments, and that reassembly in the presence of actin filaments networks spontaneously forms contractile actomyosin fibers.

Once the actomyosin complex has been formed, the subsequent activation of contraction requires release of actin-associated proteins such as caldesmon and calponin from the actomyosin

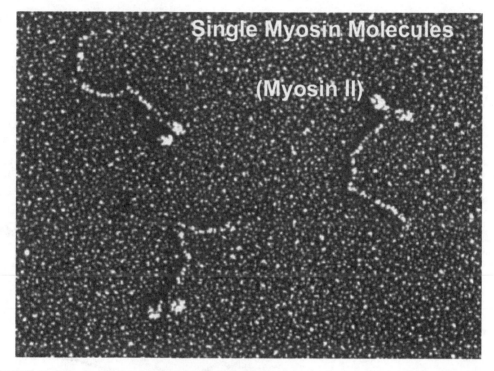

FIGURE 29: Myosin II molecules. The two globular heads on each molecule bind to actin and contain the ATPase domain. The tail forms bipolar thick filaments required for actomyosin based contraction at the rear end of migratory cells, and at the cleavage furrow. Each molecule is approximately 0.15μm long.

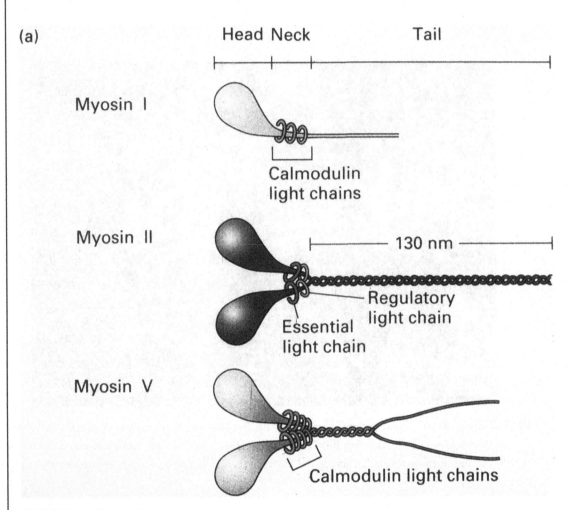

FIGURE 30: Types of myosins.

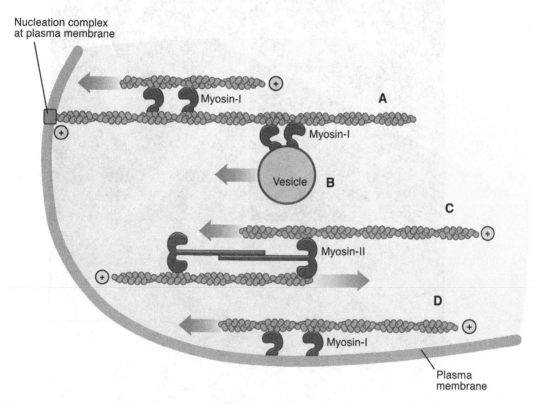

FIGURE 31: Myosin I, II movements. Based on the figure from Pollard & Earnshaw.

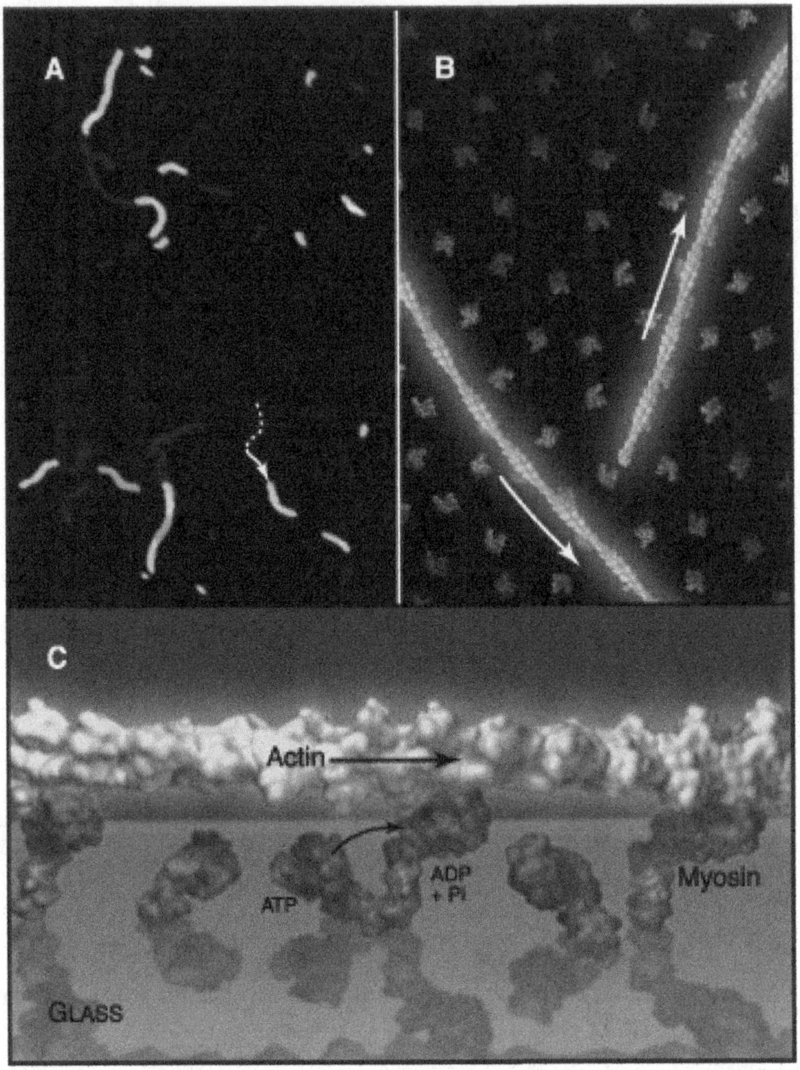

© Elsevier Ltd. Pollard & Earnshaw: Cell Biology www.studentconsult.com

FIGURE 32: In vitro motility of actin filaments gliding past myosin. (A) Actin filaments labeled with rhodamine-phalloidin to render them visible by light microscopy. (B) ATP hydrolysis by myosin moves actin filaments over the surface. (C) Drawings of actin filaments moving over myosin heads immobilized on a glass coverslip. A, Courtesy of A. Bresnick. (Reproduced from Pollard and Earnshaw: Cell Biology, www.studentconsult.com with permission of Elsevier Ltd.).

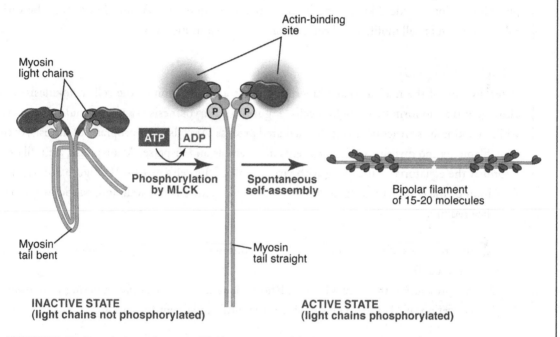

FIGURE 33: Regulation of myosin-II filament assembly by phosphorylation. (Based on the figure 16-67 from Molecular Biology of the Cell, 4th ed./Garland)

assembly in order to release the blocked interaction between the myosin head and actin filaments. These contraction regulatory proteins are, in turn, regulated by Ca^{2+}-calmodulin and phosphorylation, so that Ca^{2+} activation of calmodulin and phosphorylation of caldesmon and calponin allow an active cross-bridge to form between F-actin and myosin II heads to begin the sliding filament contraction. In summary, myosin regulation in migrating cells includes (a) phosphorylation of myosin II heavy chains or dephosphorylation of light chains to obtain motile myosin monomers, (b) myosin movement along actin filaments, (c) dephosphorylation of myosin tails or phosphorylation of myosin heads (depending on cell type) at the site of contractile fiber formation to initiate myosin thick filament assembly), (d) spontaneous formation of actomyosin contractile fibers, and (e) initiation of contraction by release of caldesmon from the actomyosin filament. Caldesmon release occurs by phosphorylation via cdc-2 kinase and binding of Ca^{2+}-calmodulin. A model depicting the various roles of myosin in cell motility and cell adhesion is shown in Figure 34.

CELL DIVISION

After division of the nucleus (karyokinesis), there is a constriction of the cell (cytokinesis) at the cleavage furrow to form two daughter cells (Figure 35). Karyokinesis is a microtubule-driven event, while cytokinesis is an actin/myosin II-mediated process. As cells enter prophase of mitosis, cortical actin filaments and myosin II thick filaments disassemble and disperse. At anaphase, actin filaments reform at the equatorial plane along with virtually all of the cell's myosin II (Figure 36). Although formation of the contractile ring and initiation of contraction is not well understood, we do know that they require:

1. microtubules of the mitotic spindle (the contractile ring falls apart if microtubules are disassembled);
2. the presence of the small GTPase, RhoA (depletion of RhoA prevents ring formation);
3. activation of MLCK to presumably initiate thick filament formation.

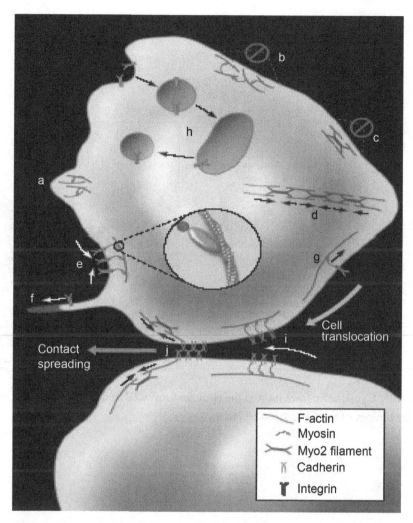

FIGURE 34: Myosin function in cell motility and cell adhesion. Lamellipodia formation and directed cell migration (a, b, c, d). Cell adhesion, receptor transport and clustering (e, g, i, j).

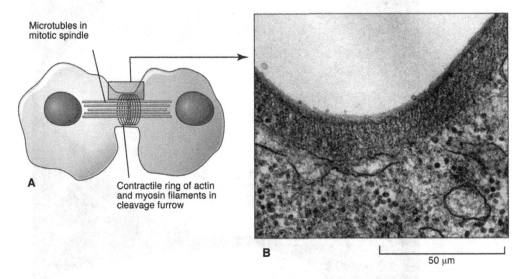

FIGURE 35: Cytokinesis: formation of the cleavage furrow. (Based on the figure 18-34 part 1 of 2 from Molecular Biology of the Cell, 4th ed./Garland)

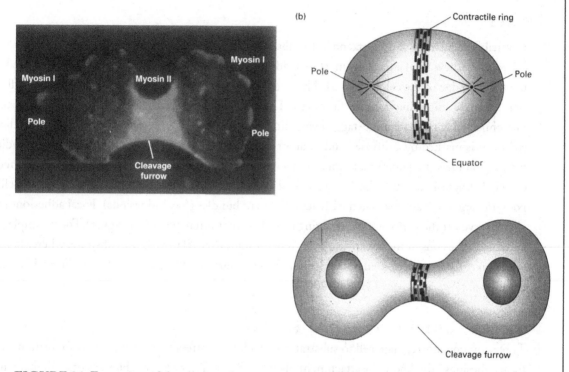

FIGURE 36: Formation of the cleavage furrow contractile ring. Courtesy of Yoshio Fukui.

The Role of Cell Adhesion in Cell Motility

Reversible regulation of cell attachment is absolutely required for migration. In order to allow the cell to move, new focal adhesions must be made at the leading edge, and old adhesions disrupted at the trailing edge of the cell (Video 3). This moving picture shows the formation of substrate adhesion sites in a migrating goldfish fibroblast. The cell was transfected with GFP-actin (green) and microinjected with rhodamine-tagged vinculin (a focal adhesion component; red). The protruding cell front is marked by a diffuse band of actin filaments (the lamellipodium), which contain radial filament bundles (filopodia) that project beyond the cell edge. Different types of adhesion foci (red) can be distinguished. Small foci are associated with lamellipodia and filopodia. Behind the lamellipodium, larger foci are associated with actin filament bundles (focal adhesions). Focal adhesions are also observed at the periphery of retracting cell edges (bottom region of the figure). Focal complexes and focal adhesions in the advancing front remain stationary relative to the substrate, whereas focal adhesions at the retracting edges can slide. (Video courtesy of Olga Krylyshkina. From J. Victor Small et al., *Nat. Rev. Mol. Cell Biol.* 2002.)

STRUCTURE OF FOCAL ADHESIONS

Three components regulate cell to substratum attachment: stress fibers, substrate composition, and focal adhesions (also known as attachment plaques). Stress fibers are large bundles of actin filaments that occupy the floor of the cell (Figure 37). In these structures, parallel arrays of F-actin are bound together with actin associated proteins such as α-actinin. The resulting actin bundle inserts into preformed focal adhesions to bind the internal cytoskeleton to the cytoplasmic face of the plasma membrane (Figure 38). An important component of the focal adhesion site is integrin. Integrins are a complex family of heterodimeric transmembrane proteins that act as plasma membrane receptors for specific substrate matrix molecules, such as fibronectin. Integrins provide the link between extracellular matrix and the internal cell cytoskeleton. A complex of different proteins is responsible for linking the cytoplasmic domain of the α,β-integrin heterodimer (also known as the fibronectin

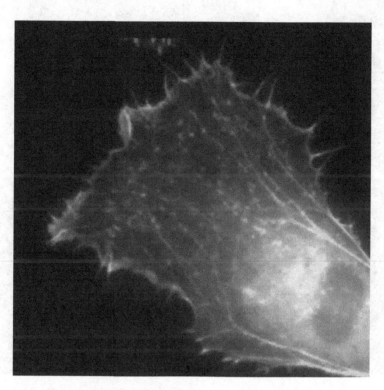

VIDEO 3: Formation of focal adhesions during cell motility. Courtesy of J. Victor Small. (See http://
www.morganclaypool.com/userimages/ContentEditor/1258743920265/Video3-fig37.mov.)

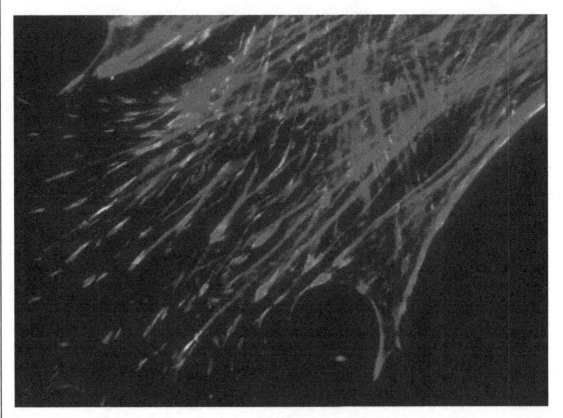

FIGURE 37: Stress fibers. F-actin bundles (red) on the cell floor of a fibroblast terminate into focal adhesion plaques (yellow spots). Newly formed focal adhesions at the leading edge of the cell are shown in green.

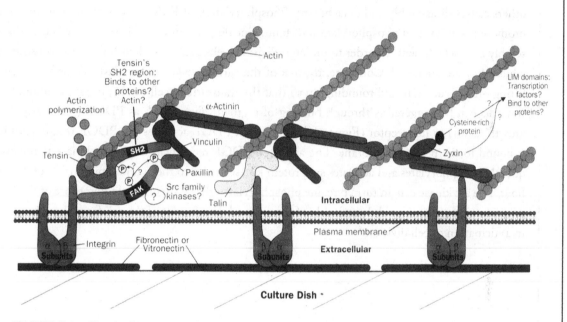

FIGURE 38: Focal adhesion site. Stress fibers bind structural proteins anchored into integrins inserted through the plasma membrane. The extracellular domain of the integrin molecule attaches to the substratum matrix.

receptor) with the stress fiber. This complex plays a crucial role in regulating attachment and detachment of the cell from its substrate.

REGULATION OF CELL ADHESION

Regulation of substrate attachment and release is mediated by tyrosine phosporylation of a number of attachment proteins, such as talin, vinculin, FAK[125], and integrin, at the focal adhesion site. Interestingly, tyrosine phosphorylation of certain attachment proteins causes assembly of the focal adhesion site, and subsequent attachment of the cell (Figure 39), while tyrosine phosphorylation of others causes disassembly and detachment. Phosphorylation of FAK[125] by rho kinase is known to promote assembly, and phosphorylation of talin, vinculin, and integrin by src kinase causes disassembly and detachment. In order to undergo division, cells must first detach from the substrate to which they are anchored. Consequently, one of the earliest stages in neoplastic transformation is lifting off the substrate and rounding up so that the transformed cell can begin proliferation. One way this can be triggered is through binding of a growth factor such as PDGF (platelet-derived growth factor) to its receptor (Figure 40). Binding of PDGF activates the PDGF receptor, which is bound to the plasma membrane. The activated PDGF receptor is a tyrosine kinase that subsequently phosphorylates and activates src protein kinase bound to the cytoplasmic surface of the cell floor. The src kinase can, in turn, tyrosine phosphorylate other proteins, including those of the focal adhesion site, causing rapid disassembly of the site and detachment of the cell from the substratum in anticipation of cell division.

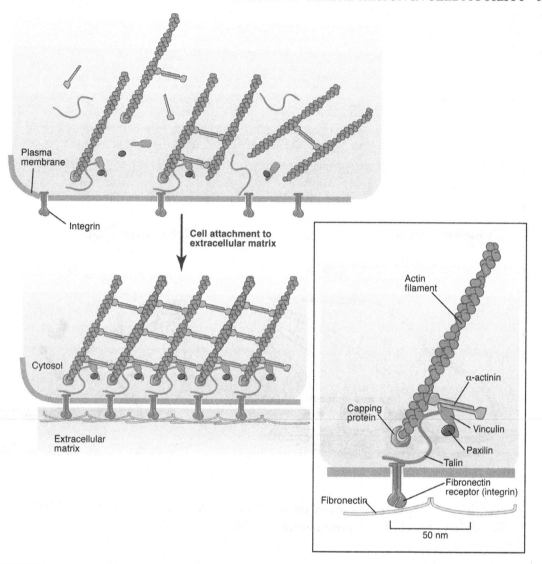

FIGURE 39: Focal adhesion site assembly. (Based on the figure 16-75 from Molecular Biology of the Cell, 3rd ed./Garland)

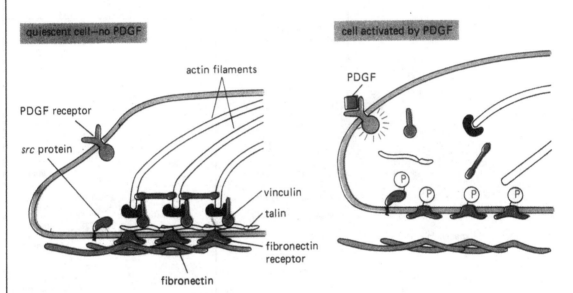

FIGURE 40: Growth factor activated cell detachment. (Reproduced based on the fig. 13-37 from Mol. Biol. of the Cell, 2nd ed. with permission of Garland).

Recommended Readings

B. Alberts, A. Johnson, J. Lewis, M. Raff, K. Roberts, and P. Walter, *Molecular Biology of the Cell*, 4th ed. Garland Science; 2002.

T. D. Pollard, W. C. Earnshaw, and J. Lippincott-Schwartz, *Cell Biology*, 2nd ed., Saunders Elsevier; 2008.